# Training Classes and the Improvement of Organizational Culture for Nurses

Dr. Platon Lomas

Kelsiehanson Verlag

# DEDICATION

To My Wonderful Readers

# CONTENTS

# ACKNOWLEDGMENTS

Thank you to everyone who strives to grow and help others grow

# 1 ABSTRACT

Background: Enhancing the culture, beliefs, and implementation of nursing regarding Evidence-Based Practices(EBP) is vital in examining the extent to which current approaches to Evidence-based practice is effective. The aim of this study was to evaluate the Effectiveness of Training classes in improving Organizational Culture, Beliefs, and Implementation of Evidence- Based Practice among nurses at Egyptian Liver Research Institute and Hospital. A quiz- experimental research design was used. This study was carried out in the Egyptian Liver Research Institute and Hospital. Study subjects included all staff nurses in aforementioned setting with a total number of 50 nurses. Data collection tools were questionnaires included the measures of organizational culture, EBP beliefs, and its implementation. Results revealed that mean score of organizational culture of evidence-based practice was significantly modified from 57.92 before the training sessions' implementation to 74.56 after implementation and 80.84 in follow up stage ($p<0.001$). Moreover, EBP Beliefs had significant improvement from 45.76 to 52.28 after implementation and reach to 55.14 in follow up stage ($p<0.001$.) Also, EBP implementation significantly enhanced from 19.50 to 26.68and more enhanced in follow up stage to 34.60.Conclusion: The training' sessions was an effective strategy in enhancing EBP culture, beliefs, and its implementation among nurses. It was recommended that the hospital should emphasis persistency of integrating EBP into daily clinical nursing care, and developing of EBP mentors and champions.

Index Terms— Evidence Based Practice,Training classes, Nurses.

# 2 INTRODUCTION

Health care is changing rapidly. The transformation from quantity- based to quality-based health care become significant; The most important element in providing the best quality- driven care is the application of evidence- based practice (Spruce, 2015). According to Mathieson, Grande,& Luker, (2018). There is an international consensus that evidence-based practice can improve outcomes for people utilizing health and social care services. for nurses, Evidence-based practice is a core requirement contributing to better clinical outcomes(Chan et al., 2020).

Evidence- based practice was defined as the ability to choose, utilize, and integrate knowledge, skills and attitudes with the intention to develop a task in a certain context. Also, EBP is the acquiring of knowledge and skills, as well as increasing positive attitudes toward EBP that will enhance a modification in behavior to implement EBP in practice(Ruzafa-Martinez et al., 2013).

The idea of EBP gathers the best accessible research evidence on one hand and clinical expertise and patients' expectations on another one, allowing for problem solving and decision making. EBP is considered as a useful tool for clinical decision making in nursing cares as it helps to reduce the health care costs by improving safety and efficiency (Panczyk & Gotlib, 2020).

Integrating EBP into healthcare organization's cultures is essential for providing safe patients' care and for promoting a prosperous culture for all nurses in the organization. Collaborating and utilizing both the clinical and the academic experts facilitates the elimination of challenges and barriers to implementing the EBP(Storey et al., 2019).

The EBP training programs demonstrate the effectiveness of improving knowledge, attitude and skills relevant to it. Moreover, These programs are effective strategies in achieving remarkable improvements in all EBP profiles' domains and skills relevant to its inquiry in nursing (El-Sayed et al., 2014)

# 3 MATERIALS AND METHODS

## Significance of the study

The nurses don't often search at the suitable information resources;they have a tendency to choose to a traditional resource such as medical mates and other medical professionals, which might don't own up-to-date or recent information, to make correct clinical decisions. Up-to-date resources as the journals and the electronic resources are used rarely. EBP training can enhance these nurses' knowledge and skillsregardingsearch operators and browsethe electronic information efficiently. These competencies promote their confidence in using the up-to-date and applicable research; additionally, they enhance sharing of the gathered information with their colleagues or with other more-experienced professionals in healthcare. In an Egyptian setting, a study of recommended that a well-constructed EBP programs should be implemented which may have a great impact on nurses knowledge about the research skills and accessibility.

## Aim of the study

This study aimed to evaluate the Effectiveness of Training classes in improving Organizational Culture, Beliefs, and Implementation of Evidence- Based Practice among nurses at Egyptian Liver Research Institute and Hospital

Research Hypothesis:

There are differences in EBP organizational culture, beliefs and implementation skills among nurses before and after implementation of the training sessions.

Subjects and method

Research Design

A quasi-experimental design was used.

Setting

This study was conducted at Egyptian Liver Research Institute and Hospital (ELRIAH) affiliated to the Association of liver Patients Care in Al-Dakahlia which is designed in the latest delightful architectural style and conforms to international standards.

The hospital consists of a ground floor, four floors, a roof,and outdoor garden. The ground floor has a large reception hall with a large number of patients. This hall is serviced by a reception and case registration office and guides the patient to the desired location. Also, it has sampling rooms, diagnostic radiology department, hospital supplies stores, kitchen, and washroom department.

The first floor contains the central laboratory of the institution and there are also a pharmacy and outpatient clinics (14 clinics that serve all specialties).Also, the first floor includes the office of the medical director of the hospital.The second floor has a highly developed endoscopy unit. Also, in this floor there are most of administrative offices, as well as the infection control unit. The third floor contains the Operating theatersand attached to it the central sterilization unit and also the intensive care as well as the inpatient department. In this floor, there is the office of the chief of nursing and the continuing training unit. The fourth and last floor includes a research unit;it also has a quality unit and the meeting rooms.

Subjects

The subjects of the study consisted of 50 nurseswho accepted to participate in the study and have more than 18 years old, they were working on the seven hospital departments: Endoscopy,Blood withdrawal, Outpatient clinics, Radiology, Internal department, Operating rooms, and Sterilization department.

Tools of Data Collection:

Three tools were used in the study.

Tool 1, Organizational Culture Evidence-based Practice scale was developed by (Fineout-overholt & Johnston, 2006)includes 18 questions regardingEBP's culture(e.g. To what extent is the nursing staff with whom you work committed to EBP,In your organization, to what extent is there a critical mass of nurses who have strong EBP knowledge and skills). Validity of this scale was supported and Cronbach alphas was>.85.

A five-pointsLikert scale, ranging from 1 = None at all, to 5 = Very much; was used to rate each statement.

Tool 2: EBP Beliefs Scale, was developed by MelnykandFineout-Overholt(2002) and modified by (Melnyk et al., 2008). Includes 16 statements about EBP's beliefs (e.g. I am clear about the steps in EBP, I am sure that I can implement EBP, I am sure that evidence-based guidelines can improve care) responses that range from 1; strongly disagree to 5; strongly agree. Scoring of thistool consists of reverse scoring; as two negatively-phrased items and summing the answers to the 16 items for a total score rangingfrom 16 to 80.Validity of this scale was supported and Cronbach alphas was>.85.

Tool 3:EBP Implementation Scale, was developed by Melnyk and Fineout-Overholt(2002) and modified by (Melnyk et al., 2008).The scale Includes 18 questions about EBP(e.g. generated a PICOT question about my practice, used evidence to change my clinical practice, evaluated the outcomes of a practice change, and shared the outcome data collected with colleagues). Nurses answering; how often, in the last 8 weeks, they have performed certain EBP initiatives, their scale ranging from 0 to 40, as "0 means no times" to 4, means ">8 times." Scoring consists of summing answersof the 18 items to get a total score, itwill range from 0 point to 72 points.The scale reliabilities greater than 0.85

In addition to the nurses'personal characteristics which included their ages, gender, their marital statuses, qualifications, and also their experience.

The training sessions

Based onresults of the assessment,evidence-based training sessions was designed to satisfy the nurses' needs.

- Knowledge test (pre-test and posttest) was developed.
- A questionnaire was developed by the researcher to measure nurses level of knowledge regarding EBP at the beginning and at the end of sessions' implementation. It consists of 20" questionrelated to content of program.
- Time schedule and teaching sessions were developed.

Field work

A written permission was secured from directors of medical and nursing department at study setting to obtain their approval. On data collection, a verbal agreement was taken from every nurse participated in the study, after a clear and simple explanation of the purpose and importance of the study.

The intended tools of data collection were translated from English to Arabic language and retranslated and revised by panel of experts.

A pilot study was conducted on10% of nurses from the studied hospital that fulfilled the study'scriteria then was excluded from the original study.

Individual interview was carried out to assess the nurses' perception of the organizational culture,beliefs, and the implementation of EBP. Each interview took a period of about 25-30 minutes.

Development of the training sessions: Based on the results of the pre assessment the time management training program was developed as well as the time schedules and teaching sessions.

Before application of the training program for nurses, permission was taken from the hospital directors and the nursing service directors.

Knowledge test was administered to the nurses before the beginning of the program to collect data related to their level of knowledge; the test was completed in 30 minutes.

The training classeswere implemented for nurses at the studied hospital. The following teaching methods were utilized: interactive lecture, discussion, brainstorming, and work in small groups. The intervention was five training sessions consisted of five theoretical lectures; eachof themwas followed with a practice session. The training sessions covered around 25 hours of training (a total of 10 hrs. of theoretical lectures and 15 hours of practice) and three consecutive days (February 25-28).

The knowledge test that was used at the beginning of the program was given again to nurses at the end of the program; to evaluate the gained knowledge in comparison with the pre-test.

Assessment tools: (organizational culture scale, EBP Beliefs Scale, and the EBP Implementation Scale) that were used in the pre- assessment were used immediately at the end of the program and after 3 months of the program implementation, to reveal the change in nurses' perception,behaviors and outcome in relation to EBP.Practice activities includedspecific clinical questions, using the Patient, Intervention, Comparison, Outcome, and the Timeframes(PICOT) formats, also developing a search strategy to identify the appropriate keywords and the most applicable database then conducting the search processbythe provided computers in the labs, andfinally conducting a critical appraisal on the evidence resulted and answering the clinical questions.

The Media of teaching involvedPowerPoint slides, flipped charts, and handouts. The available English databases included the MD Consult, Cumulative Index to Nursing & Allied Health Literature (CINAHL), MEDLINE, the Cochrane Library, ProQuest, Micromedex, UpToDate, and also the DynaMed. They were selected owing to their popularity.

Ethical Consideration

An oral informed consent was obtained from each participant. All of themhad the rights to refuse to participate or withdraw from the study at any time with no consequences. Additionally, they were informed about their anonymity and collected information'sconfidentiality, and it would be used to fulfill the aim of this scientific study.

Statistical analysis of the data:

Data were fed to the computer and analyzed using IBM SPSS software package version 20.0. (Armonk, NY: IBM Corp) Qualitative data were described using number and percent.Quantitative data were described using, mean, standard deviation, Significance of the obtained results was judged at the 5% level.

The used tests were

1- ANOVA with repeated measures

For the normally-distributed quantitative variables, to compare between more than two periods or stages, and Post Hoc test (Bonferroni adjusted) for pairwise comparisons

2- Pearson's coefficient test

To correlate between two normally-distributed quantitative variables

3- Student t-test

For normally distributed quantitative variables, to compare between two studied groups

4- F-test (ANOVA)

For normally-distributed quantitative variables, to compare between more than two groups

3. Results

**3. RESULTS AND DISCUSSIONS:**

Table (1) Personnel characteristics of the studied nurses (n = 50)

| Personnel characteristics | No. | % |
|---|---|---|
| Gender | | |
| Male | 28 | 56.0 |
| Female | 22 | 44.0 |
| Age | | |
| < 20 | 2 | 4.0 |
| 20 –< 30 | 38 | 76.0 |
| ≥ 30 | 10 | 20.0 |
| Min. – Max. | 18.0 – 40.0 | |
| Mean ± SD. | 25.28 ± 4.92 | |
| Qualifications | | |
| Nursing diploma | 4 | 8.0 |
| Diploma of Nursing institute | 33 | 66.0 |
| Bachelor of Nursing | 10 | 20.0 |
| Higher studies | 3 | 6.0 |
| Experience | | |
| <6 month | 13 | 26.0 |
| 6 - <12 | 3 | 6.0 |
| 12 - <18. | 5 | 10.0 |
| 18 - <24 | 2 | 4.0 |
| 24- <30 | 8 | 16.0 |
| 36+ | 19 | 38.0 |
| Min. – Max. | 0.0 – 84.0 | |
| Mean ± SD. | 26.54 ± 23.80 | |
| Have you studied evidence-based practice in your courses? | | |
| No | 33 | 66.0 |
| Yes | 17 | 34.0 |
| Have you attended evidence-based practice training courses? | | |
| No | 35 | 70.0 |
| Yes | 15 | 30.0 |

The study involved 50 nurses; their age ranged between 18 and 40 years with mean age of 25.28 ± 4.92.Male nurses represent 65% of the study sample as shown in table 1. The majority of them had diploma of nursing institute with mean years of experience 26.54 ± 23.80. The great majority of studied nurses hadn't study evidence-based practice in their courses. However, only (30%) of them had attended evidence-based practice training courses.

Table (2): Comparison between the different periods according to scores

| | Pre (n = 50) | Post 3 (n = 50) | Post 6 (n = 50) | F | P |
|---|---|---|---|---|---|
| | Mean ± SD | Mean ± SD | Mean ± SD | | |
| Organizational Culture Evidence-based Practice | | | | | |
| Total score | 57.92 ± 20.32 | 74.56 ± 7.15 | 80.84 ± 4.73 | 74.337* | <0.001* |
| % score | 60.33 ± 21.16 | 77.67 ± 7.45 | 84.21 ± 4.93 | | |
| Sig. bet. groups | p1<0.001*, p2<0.001*, p3<0.001* | | | | |
| EBP Beliefs Scale | | | | | |
| Total score | 45.76 ± 10.61 | 52.28 ± 4.98 | 55.14 ± 4.43 | 69.703* | <0.001* |
| % score | 71.50 ± 16.59 | 81.69 ± 7.78 | 86.16 ± 6.92 | | |
| Sig. bet. groups | p1<0.001*, p2<0.001*, p3<0.001* | | | | |
| EBP Implementation Scale | | | | | |
| Total score | 19.50 ± 17.24 | 26.68 ± 13.75 | 34.60 ± 10.35 | 133.801* | <0.001* |
| % score | 27.08 ± 23.94 | 37.06 ± 19.10 | 48.06 ± 14.37 | | |
| Sig. bet. groups | p1<0.001*, p2<0.001*, p3<0.001* | | | | |

F: F test (ANOVA) with repeated measures, Sig. bet. Periods was done using Post Hoc Test (Bonferroni)

p: p value for comparing between different periods

p1: p value for comparing between pre and post 3

p2: p value for comparing between pre and post 6

p3: p value for comparing between post 3 and post 6

*: Statistically significant at $p \leq 0.05$

Table (2) shows the evidence-based practice organizational culture, EBP beliefs, and EBP implementation among the studied nurses at different training classes' stages. Regarding organizational culture of evidence-based practice the mean score was significantly modified from57.92 ± 20.32 mean score before the training classes' implementation to 74.56 ± 7.15 after implementation and 80.84 ± 4.73 in follow up stage (p<0.001). Moreover, EBP Beliefs had significant improvement from the mean score 45.76 ± 10.61before the classes to be 52.28 ± 4.98 after implementation and reach to 55.14 ± 4.43 in follow up stage (p<0.001.) Also, EBP implementation significantly enhanced from the mean score 19.50 ± 17.24 to 26.68 ± 13.75 and more enhanced in follow up stageto 34.60 ± 10.35

Table (3): Correlation between different parameters

| | | Pre | Post 3 | Post 6 |
|---|---|---|---|---|
| Organizational Culturevs EBP Beliefs | r | 0.627* | 0.545* | 0.352* |
| | p | <0.001* | <0.001* | 0.012* |
| Organizational Culturevs EBP Implementation | r | 0.174 | 0.032 | 0.214 |
| | p | 0.226 | 0.826 | 0.136 |
| EBP Beliefsvs EBP Implementation | r | -0.042 | -0.003 | 0.024 |
| | P | 0.774 | 0.984 | 0.869 |

r: Pearson coefficient

*: Statistically significant at $p \leq 0.05$

Regarding correlation between EBP organizational culture, EBP Beliefs, and EBP Implementation table 3 demonstrates a statistically significant positive correlation between organizational culture of EBPand EBP Beliefs before and after implementation of training classes.

Table (4): Comparison between pre and post according to overall knowledge

| Overall knowledge | Pre | Post | t | P |
|---|---|---|---|---|
| Total score | | | 24.946* | <0.001* |
| Min. – Max. | 0.0 – 8.0 | 8.0 – 17.0 | | |
| Mean ± SD. | 4.14 ± 1.81 | 13.44 ± 2.27 | | |
| % score | | | | |
| Min. – Max. | 0.0 – 40.0 | 40.0 – 85.0 | | |
| Mean ± SD. | 20.70 ± 9.04 | 67.20 ± 11.35 | | |

t: Paired t-test

p: p value for comparing between different periods

*: Statistically significant at $p \leq 0.05$

As evident in table (4) the total means score of studied nurses overall knowledge regarding evidence based practice were 4.14 ± 1.81 before training classes' implementation and significantly increased to13.44 ± 2.27 after implementation(p<0.001).

Table (5): Relations between organizational culture and personnel characteristics at the different periods

| Personnel characteristics | pre | post | Follow-up |
|---|---|---|---|
| | Mean ± SD. | Mean ± SD. | Mean ± SD. |
| Gender | | | |
| Male | 54.91 ± 23.75 | 76.12 ± 6.92 | 83.15 ± 4.61 |
| Female | 67.23 ± 15.20 | 79.64 ± 7.80 | 85.56 ± 5.10 |
| t(p) | 2.226*(0.031*) | 1.691(0.097) | 1.753(0.086) |
| Age | | | |
| <20 | 79.69 ± 5.16 | 88.02 ± 6.63 | 90.10 ± 5.16 |
| 20 –< 30 | 62.23 ± 20.88 | 78.21 ± 7.07 | 84.65 ± 4.85 |
| ≥ 30 | 49.27 ± 20.20 | 73.54 ± 6.91 | 81.35 ± 3.85 |
| F(p) | 2.498(0.093) | 3.998*(0.025*) | 3.607*(0.035*) |
| Department | | | |
| Endoscopy | 61.94 ± 27.01 | 82.13 ± 7.21 | 87.10 ± 5.08 |
| Blood withdrawal | 71.88 ± 6.83 | 80.56 ± 4.34 | 84.37 ± 4.77 |
| Outpatient clinics | 72.22 ± 4.92 | 80.90 ± 5.14 | 85.42 ± 4.77 |
| Radiology | 52.18 ± 26.67 | 75.95 ± 8.76 | 83.33 ± 5.95 |
| Internal department | 72.14 ± 8.76 | 78.91 ± 6.65 | 84.37 ± 3.94 |
| Operating rooms | 52.08 ± 13.87 | 72.16 ± 4.35 | 81.63 ± 3.27 |
| Sterilization department | 55.21 | 70.83 | 79.17 |
| F(p) | 1.350(0.257) | 2.680*(0.027*) | 1.629(0.162) |
| Qualifications | | | |
| Nursing diploma | 72.92 ± 4.74 | 81.77 ± 3.12 | 85.16 ± 2.15 |
| Nursing institute | 63.92 ± 19.89 | 78.76 ± 7.81 | 84.94 ± 5.35 |
| Bachelor of Nursing | 54.17 ± 20.86 | 75.10 ± 5.80 | 82.71 ± 3.84 |
| Higher studies | 24.65 ± 5.14 | 68.75 ± 3.76 | 79.86 ± 3.66 |
| F(p) | 4.830*(0.005*) | 2.725(0.055) | 1.416 (0.250) |
| Have you studied evidence-based practice in your courses? | | | |
| No | 56.38 ± 23.67 | 77.15 ± 7.58 | 84.19 ± 4.92 |
| Yes | 68.01 ± 12.50 | 78.68 ± 7.32 | 84.25 ± 5.10 |
| t(p) | 2.275*(0.027*) | 0.684 (0.497) | 0.045 (0.964) |

t: Student t-test F: F for ANOVA test

p: p value for association between different categories

*: Statistically significant at $p \leq 0.05$

Regarding the relations between organizational culture and nurses' personal characteristics at the training classes' stages, Table 5 indicates that nurses gender had statistically significant relation with nurses EBP organizational culture before the training classes implementation (p=0.031) as female nurses had higher mean score than male. Moreover, nurses' age was positively correlated with their organizational culture in the post and follow-up stage of training; where younger nurses had the higher mean scores than other nurses. Also, work department were positively correlated with organizational culture (p=0.027), as nurses who working in Endoscopy department had higher mean score post training. Additionally, the table points a statistically significant relation between organizational culture and diploma nurses (p=0.005) and those who had study evidence-based practice in their courses (p= 0.027) before training classes implementation.

Table (6): Relations between EBP beliefs and personnel characteristics at the different periods

| Personnel characteristics | pre | post | Follow-up |
|---|---|---|---|
| | Mean ± SD. | Mean ± SD. | Mean ± SD. |
| Department | | | |
| Endoscopy | 71.51 ± 19.20 | 82.69 ± 7.72 | 87.02 ± 6.13 |
| Blood withdrawal | 85.94 ± 5.41 | 89.06 ± 1.56 | 92.19 ± 0.00 |
| Outpatient clinics | 89.06 ± 0.00 | 90.10 ± 0.90 | 93.75 ± 3.13 |
| Radiology | 76.56 ± 12.75 | 83.38 ± 6.38 | 88.35 ± 6.19 |
| Internal department | 74.22 ± 9.49 | 82.03 ± 5.54 | 86.13 ± 6.27 |
| Operating rooms | 58.10 ± 14.79 | 75.14 ± 7.98 | 80.26 ± 6.02 |
| Sterilization department | 45.31 | 71.88 | 75.0 |
| F(p) | 3.656*(0.005*) | 3.579*(0.006*) | 4.095*(0.002*) |
| Experience | | | |
| <6 month | 69.11 ± 13.59 | 80.89 ± 5.78 | 84.62 ± 6.87 |
| 6 - <12 | 60.42 ± 19.78 | 79.69 ± 8.70 | 81.77 ± 8.88 |
| 12 - <18. | 63.75 ± 17.20 | 79.69 ± 9.11 | 85.00 ± 6.77 |
| 18 - <24 | 75.78 ± 3.31 | 79.69 ± 2.21 | 84.38 ± 0.00 |
| 24- <30 | 87.89 ± 3.42 | 89.26 ± 1.76 | 92.77 ± 2.89 |
| 36+ | 69.57 ± 18.56 | 80.10 ± 9.14 | 85.61 ± 7.15 |
| F(p) | 2.519*(0.043*) | 2.022 (0.094) | 2.135 (0.079) |

t: Student t-test F: F for ANOVA test

p: p value for association between different categories

*: Statistically significant at $p \leq 0.05$

As Table 6 describes, only work department and experience had a correlation with beliefs of EBP; a statistically positive significant relations could be identified between nurses' working department and beliefs of EBP where the mean scores were higher among nurses who working in outpatients' clinics all over the training classes' stages. Moreover, only statistically positive significant relation was found between nurses' years of experience and their beliefs before implementation.

Table (7): Relations between EBP implementation and personnel characteristics at the different periods

| Personnel characteristics | pre | post | Follow-up |
|---|---|---|---|
| | Mean ± SD. | Mean ± SD. | Mean ± SD. |
| Department | | | |
| Endoscopy | 14.96 ± 19.94 | 28.31 ± 15.85 | 45.19 ± 12.22 |
| Blood withdrawal | 9.26 ± 3.50 | 25.93 ± 4.88 | 38.43 ± 5.78 |
| Outpatient clinics | 11.11 ± 3.67 | 23.61 ± 5.01 | 34.26 ± 8.93 |
| Radiology | 42.42 ± 31.18 | 47.98 ± 26.19 | 57.32 ± 18.79 |
| Internal department | 32.29 ± 20.53 | 39.41 ± 17.50 | 46.01 ± 14.78 |
| Operating rooms | 29.67 ± 19.19 | 40.03 ± 14.63 | 49.37 ± 10.14 |
| Sterilization department | 47.22 | 52.78 | 55.56 |
| F(p) | 2.341*(0.048*) | 1.814(0.119) | 1.776(0.127) |
| Have you studied evidence-based practice in your courses? | | | |
| No | 17.97 ± 17.52 | 30.13 ± 14.27 | 44.02 ± 10.88 |
| Yes | 44.77 ± 25.26 | 50.49 ± 20.47 | 55.88 ± 17.23 |
| t(p) | 4.395*(<0.001*) | 3.667* (0.001*) | 2.585* (0.017*) |
| Have you attended evidence-based practice training courses? | | | |
| No | 17.98 ± 16.79 | 29.80 ± 13.31 | 43.17 ± 10.18 |
| Yes | 48.33 ± 25.11 | 53.98 ± 20.18 | 59.44 ± 16.50 |
| t(p) | 5.023*(<0.001*) | 4.260* (<0.001*) | 3.541* (0.002*) |

t: Student t-test F: F for ANOVA test

p: p value for association between different categories

*: Statistically significant at p ≤ 0.05

Regarding the relations between nurses' EBP implementation and their personal characteristics, Table 7 indicates a statistically positive significant relations with their working department (p=0.048) as those working in radiology department had the highest mean score. Also, it can be noticed that those who had study and training in EBP had higher mean scores of implementing EBP among the training classes' stages.

I. Discussion

Evidence-based practice is the highest standard of nursing practice and nursing education; it is comprised of at least three overlapping factors. First, the best available scientific evidence is based on findings of well-conducted studies. Second, interventions are based on the opinion of expert nurses. Third, patient preferences are based on patient's culture and the situational circumstances (Rubin, 2008 and Wallen et al., 2010).

Educational program on nurses' Evidence-based practice may enhance research utilization in clinical practice, prepare nurses to use evidence on an organization wide level, and affect nurses' beliefs about and attitudes toward EBP (Fennel, 2013, Linde, 2015).The results of a study carried byAlmaskari, (2017) provided evidence for recommendations to integrate EBP in educational programs and clinical settings, as well as recommendations for future research.

The current study's aim was to investigate the effect of structured training classes on EBP culture, beliefs and implementation skills; among nurses at Egyptian Liver Research Institute and Hospital.The findings point to generally improvement of nurses' organizational culture, beliefs and implementation of EBP after training classes' implementation.

The study sample included nurses with a relatively early and wide age spectrum, from the beginning of second to the fourth decades of age, thus representing a good mix of different generations within a young age. This was important since the readiness to learn and acquiring Knowledge and skills is undoubtedly influenced by age. They also had a wide span of experience in the hospital, which certainly have an impact on their performance level.

Patelarou et al., (2013); stated that the youngest nurses with the fewest years in practice possessed more positive responsesregardingimplementation of EBP and readiness of the organization. Farokhzadian et al., (2015) highlighted that young nurses were more motivated to use a variability of search features when compared to their seniors. This may be asthey are more aware and used to use computers and browse the internet.

Meijers et al. (2006) suggested that providing education and time for nurses to engage in applied research activities can influence their beliefs toward EBP, and competency building in EBP implementation. Wan, (2017) added that, The EBP educational intervention significantly improved nurses' beliefs about EBP, knowledge of EBP, and their perceived ability to implement EBP.

According to the present study findings, nurses had a significant higher increase in the scores of perceived organizational cultures and EBP's belief and implementation than before training classes' implementation.

These findings could be related to the organizational culture and climate, the multi-facetedmeans of support, the time needed for staff to become involved in research activities, and the educationprovision to them, empowerment, and work engagement as well as, the role of a nurse as a cornerstone in care, and feasibility of access to resources. Also,the nurses' persistency to attend the training' classes and improve their roles and quality of performance.

In the systematic review byEstabrooks, Floyd, Scott-Findlay, O'Leary, and Gushta (2003), they clarified that nurses' culturesand beliefs toward EBP were correlated with the utilization of evidence in practice. Likewise, Wallin, Bostrom, Wikblad, and Ewald (2003) demonstrated that nurses who received EBP training were more persistent in promoting evidence-based nursing.

A study of Hellier, (2015) supporting the previous findings,reported that participants were overwhelming supportive of EBP. Furthermore, participants showedthat they had high ability to judge, understand, and integrate evidence into practice. Likewise, it was also detected that participating in the professional training programs increased their usage of several searching features.

Similar concerns regarding the consequences of EBPwere reported by nurses in a study carried out in Egyptby Shazly et al., (2018) and demonstrated that there were positive beliefs among nurses concerningthe EBP. The majority of them possessedpositive beliefs regarding applying the EBP in clinical settings and they also approved that; the applying EBP is very necessary in all nursing care; and the research findingsareessential in day-to-day practice. Also, they claimed that EBP improves the qualities of patient's care.

Furthermore, the present study findings demonstrated a positive correlation between the three parameters of the study.It was clear that organizational culture was closely related to EBP beliefs before and after implementing of training classes.This finding can be interpreted as Nurses try to adapt their interventions to different patient preferences and clinical situations while utilizing the best practices based on the findings of updated available studies.

Effective EBP education is immensely in need for improving nurses 'knowledge of EBP, beliefs about EBP, attitudes toward EBP, and their perceived abilities to implement EBP (Dizon et al., 2014; Mollon et al., 2012 Parrish & Rubin, 2011).

As regard nurses' knowledge about Evidence-based practice. The pre-post multiple choice question tests clarified that the total mean score of studied nurses overall knowledge regarding evidence based practice were significantly increased after training implementation;as educational interventionis considered to enhancenurses' knowledge about EBP.

Incongruence with the foregoing present study findings, Hart et al., 2008&Mollon et al., 2012 reported that Structured EBP educational interventions may improve nurses' knowledge and skills of EBP, their attitudes toward EBP, and their perceived ability to implement EBP.Stillwell (2010) &Black et al., (2015) emphasized that practicing nurses, required education on EBP so that they may become knowledgeable and competent in the skills of using EBP in nursing.

Regarding the relations between study parameters and nurses' personal characteristics, it was evident that nurses' gender had only a statistical significant relation with nurses EBP organizational culture before the training classes' implementation as female nurses had higher mean score than male. Moreover, nurses' age was positively correlated with their organizational culture in the post and follow-up stage of training; where younger nurses had the higher mean scores than other nurses. Also, work department were positively correlated with organizational culture post training and with EBP beliefs in the three training stages also positively correlated with EBP implementation in the pre trainingstage. Additionally, a statistical significant relationwas found between organizational culture and diploma nurses before training implementation.While onlyexperience had a correlation with beliefs of EBP. Also, it can be noticed that those who had study and training in EBP had higher scores of implementing EBP among the training stages.

Finally,Nurses must accept that EBP is an ongoing process and lifelong learning process. Nurses must adapt their interventions to different patient preferences and clinical situations while using theEffective educational interventions to equip nurses with knowledge and skills to becomeEBP competent. (Theriaque & Whitehead, 2018)

Conclusion

Findings indicated that the EBP Programincluded a series of training classes with enduring skills and building activities have great and positive effects on nurses' perception onthe organizational cultures, their EBP beliefs and its implementation.

## Recommendations

In view of the main study findings, the health care facility should:

- Emphasis the Persistency of the integration of evidence-based practice into daily clinical nursing care, and covering the gap between a scientific researches and nursing care.
- Empower the culture of EBP and the availability of resources for continuity of implementation.
- Development of evidence-based practice champions and mentors through training programs to maintain the culture of evidence-based practice.
- Facilitating accessibility to reference library settings, computer and internet labs to promote the online search of EBP and maintain a reward system for motivating the innovation of nursing staff.

.

References


[1] Black, A., Balneaves, L. G., Garossino, C., Puyat, J. H., & Qian, H. (2015). Promoting evidence-based practice through a research training program for point-of-care clinicians. Journal of Nursin Administration, 45(1), 14-20. doi: 10.1097/ NNA.0000000000000151

[2] Chan, E.-Y., Glass, G. F., & Ni Phang, K. (2020). Evaluation of a hospital-based nursing research and evidence-based practice mentorship program on improving nurses' knowledge, attitudes, and evidence-based practice. Journal of Continuing Education in Nursing, 51(1), 46–52. https://doi.org/10.3928/00220124-20191217-09

[3] El-Sayed, S. H., Hassona, F. M., & Winkelman, C. (2014). Effect of a Training Program on Evidence Based Practice Profiles and Skills among Master Nursing Students. Zagazig Nursing Journal, 10(1), 1–18. https://doi.org/10.12816/0029300

[4] Estabrooks C.A., Floyd J.A., Scott-Findlay S., O'Leary K.A. & Gushta M. (2003). Individual determinants of re- search utilization: A systematic review. Journal of Ad- vanced Nursing, 43, 506–520.

[5] Farokhzadian, J., Khajouei, R., & Ahmadian, L. (2015). Information seeking and retrieval skills of nurses: Nurses readiness for evidence based practice in hospitals of a medical university in Iran. International Journal of Medical Informatics, 84(8), 570–577. https://doi.org/10.1016/j.ijmedinf.2015.03.008

[6] Fineout-overholt, S. E. E., & Johnston, L. (2006). Teaching EBP: Implementation of evidence: Moving from evidence to action. Worldviews on Evidence-Based Nursing, 3(4), 194–200. https://doi.org/10.1111/j.1741-6787.2006.00070.x

[7] Hart, P., Eaton, L., Buckner, M., Morrow, B. N., Barrett, D. T., Fraser, D. D.,…Sharrer, R. (2008). Effectiveness of a computer-based educational program on nurses' knowledge, attitude, and skill level related to evidence-based practice. Worldviews on Evidence-Based Nursing, 5(2), 75-84. doi:10.1111/j.1741- 6787.2008.00123.x

[8] Hellier I. (2015).Factors Affecting Implementation Of Evidence-Based Practice Among Nurse Practitioners, A Dissertation Submitted to the School of Graduate Studies and Research in Partial Fulfillment of the Requirements for the Degree Doctor of Philosophy (Issue December).
[9] Mathieson, A., Grande, G., & Luker, K. (2018). Strategies, facilitators and barriers to implementation of evidence-based practice in community nursing: A systematic mixed-studies review and qualitative synthesis. Primary Health Care Research and Development, 20. https://doi.org/10.1017/S1463423618000488
[10] Melnyk, B. M., Fineout-Overholt, E., Feinstein, N. F., Sadler, L. S., & Green-Hernandez, C. (2008). Nurse Practitioner Educators' Perceived Knowledge, Beliefs, and Teaching Strategies Regarding Evidence-Based Practice: Implications for Accelerating the Integration of Evidence-Based Practice Into Graduate Programs. Journal of Professional Nursing. https://doi.org/10.1016/j.profnurs.2007.06.023
[11] Melnyk B. & Fineout-Overholt E. (2002). Putting research into practice. Reflections on Nursing Leadership, 28(2), 22–25.
[12] Meijers, J. M. M., Janssen, M. A. P., Cummings, G. G., Wallin, L., Estabrooks, C. A., & Halfens, R. Y. G. (2006). Assessing the relationships between contextual factors and research utilization in nursing: Systematic literature review. Journal of Advanced Nursing, 55(5), 622-635. doi:10.1111/j.1365-2648.2006.03954.x
[13] Mokhtar, S. Majid, S. Foo, X. Zhang, Y.-L. Theng, Y.-K. Chang, et al., Evidence-based practice and related information literacy skills of nurses in Singapore: an exploratory case study, Health Inform. J. 18 (2012) 12–25
[14] Mollon, D., Fields, W., Gallo, A. M., Wagener, R., Soucy, J., Gustafson, B., & Kim, S. C. (2012). Staff practice, attitudes, and knowledge/skills regarding evidence- based practice before and after an educational intervention. Journal of Continuing Education in Nursing, 43(9), 411-419. doi:10.3928/00220124-20120716-89
[15] Panczyk, M., & Gotlib, J. (2020). EBP e-Toolkit: Providing a Teaching and Learning Open and Innovative Toolkit for Evidence-Based Practice to Nursing European Curriculum - The second transnational project meeting in Warsaw. Pielegniarstwo XXI Wieku, 18(3), 197–199. https://doi.org/10.2478/pielxxiw-2019-0023
[16] Rubin, A. (2008). Practitioner's guide for using research for evidence-based practice. Hoboken, NJ: John Wiley & Sons.
[17] Ruzafa-Martinez, M., Lopez-Iborra, L., Moreno-Casbas, T., & Madrigal-Torres, M. (2013). Development and validation of the competence in evidence based practice questionnaire (EBP-COQ) among nursing students. BMC Medical Education, 13(1). https://doi.org/10.1186/1472-6920-13-19
[18] Shazly, M., Abdel- Azeem, H., & Mohamed Ahmed, N. (2018). Nurses'''' Beliefs and Barriers toward Evidence – Based Practice. Egyptian Journal of Health Care, 9(2), 87–105. https://doi.org/10.21608/ejhc.2018.10685
[19] Spruce, L. (2015). Back to basics: Implementing evidence-based practice. AORN Journal, 101(1), 106-114.e4. https://doi.org/10.1016/j.aorn.2014.08.009

[20] Stillwell, S. B. (2010). National survey shows the majority of nurses use very little research in the first 2 years after their graduation, highlighting a gap between research and clinical practice. Evidence-Based Nursing, 13(4), 104. doi:10.1136/ ebn13.4.104
[21] Storey, S., Wagnes, L., Lamothe, J., Pittman, J., Cohee, A., & Newhouse, R. (2019). Building evidence-based nursing practice capacity in a large statewide health system: A multimodal approach. Journal of Nursing Administration, 49(4), 208–214. https://doi.org/10.1097/NNA.0000000000000739
[22] Theriaque, T., & Whitehead, D. (2018). Educational Training of Staff Nurses for Evidence-Based Practice. ProQuest Dissertations and Theses, 81. http://ezproxy.puc.cl/docview/2038441688?accountid=16788%0Ahttp://todosibuc.uc.cl/openurl/56PUC_INST/56PUC_INST_SP?url_ver=Z39.88-2004&rft_val_fmt=info:ofi/fmt:kev:mtx:dissertation&genre=dissertations+%26+theses&sid=ProQ:ProQuest+Dissertations+%26+Theses+
[23] Athina E. Patelarou ,EvridikiPatelarou , Hero Brokalaki , VassilisDafermos , Linda Thiel , Christos D. (2013). Current Evidence on the Attitudes, Knowledge and Perceptions of Nurses Regarding Evidence-Based Practice Implementation in European Community Settings: A Systematic Review Pages 230-244 | Published online: 12 Nov
[24] Wallin, L., Bostrom, A. M., Wikblad, K., & Ewald, U. (2003). Sustainability in changing clinical practice promotes evidence-based nursing care. Journal of Advanced Nursing, 41(5), 509-518. doi:10.1046/j.1365-2648.2003.02574x

Made in United States
North Haven, CT
04 November 2026

21743908R00017